Nolis Bracho Moran
José Labrador Ramírez
José Ramón Vielma Guevara

BIOFUELS IN VENEZUELA

Nolis Bracho Moran
José Labrador Ramírez
José Ramón Vielma Guevara

BIOFUELS IN VENEZUELA

A SOME LOOK AT ETHANOL PRODUCTION PERFORMANCE IN SUGAR CANE (Saccharum spp. hybrid) CROPS

Authors:

- **Nolis de Jesús Bracho Morán**. Agricultural Production Engineer from the Universidad Nacional Experimental Sur del Lago "Jesús María Semprum" UNESUR. *Magister Scientiae* in Agronomy, mention in Plant Production, Universidad Nacional Experimental del Táchira. Assistant Professor at UNESUR, Santa Bárbara de Zulia, Zulia State.

- **José Rafael Labrador Ramírez.** Engineer and *Magister Scientiae* in Agronomy. Professor at the Universidad Nacional Experimental Sur del Lago "Jesús María Semprum" UNESUR, Santa Bárbara de Zulia, Zulia State.

- **José Ramón Vielma Guevara.** Graduate in Bioanalysis, *Magister Scientiae* in Cellular Biology at the Universidad de Los Andes, Mérida. Teaching Component and University Teaching at the Universidad del Zulia, Maracaibo. Instructor at the Universidad Politécnica Territorial "José Félix Ribas", Barinas, Barinas State, Venezuela.

General index

Introduction..5

Chapter I. The need for biofuels development in Venezuela: an oil-producing country seeking export diversity 7

Chapter II. Literature review on the theoretical basis and background to the development of this research 14

Chapter III. Methodological framework 29

Chapter IV. Results and discussion... 36

Chapter V. Conclusions and Recommendations 45

Foreword

Venezuela is an oil-exporting country and a member of the Organisation of Petroleum Exporting Countries (OPEC) since its creation in 1960 in the city of Baghdad in Iraq, with its headquarters in Vienna, the capital of Austria. Traditionally, we have been a mono-producer of this mineral resource, and we have the largest amount of proven crude oil reserves in the world.

However, this enormous wealth of our subsoil is historically at odds with the great inequality in the distribution of wealth among the population, where overcrowding, misery, hunger, lack of employment hit a large part of our population hard, making us very vulnerable in terms of health and nutrition in the female population and especially in the case of young children. In addition to the above, it is generally women who are the "heads" of the family, who are unable to access formal or technical education, nor do they have access to well-paid jobs.

We are considered a developing country, which is why we are at this historic stage in the immediate need to diversify our production of goods, services and food in order to guarantee social equity. That is why our largest state company, Petróleos de Venezuela, set its sights on the development of biofuels some time ago, with ethanol (CH_3CH_2OH) in its sights, following the example of our sister country Brazil, which is the world's largest exporter of this product, based on sugar cane. This is why we dedicate this book to our practical experience with the production of ethanol from sugar cane cultivars in the region south of Lake Maracaibo in Venezuela.

Summary

In order to evaluate the yield of sugar cane (*Saccharum* spp. hybrid) for ethanol production, eleven cultivars were evaluated: V91-8, V98-86, V91-01, V99-271, C323-368, V98-120, V99-236, B80-408, V00-50, V99-190, CP74-2005 during one production cycle. The trial was carried out in the field of the Universidad Nacional Experimental Sur del Lago "Jesús María Semprum" (UNESUR), in a randomised block design, choosing plants from the central furrow of each experimental plot (45 m2/treatment). The criteria evaluated were: production in tonnes of cane per hectare (TCH), tonnes of sugar per hectare (TAH), percentage of polysaccharides (% POL), litres of ethanol per hectare (LtEt/hA), efficiency (LtEt/TC) and concentration of ethanol produced. The results indicate non-significant differences between treatments by cultivar effect for the variable TAH, and highly significant for efficiency (LtEt/TC) ($p \geq 0.001$) and ethanol concentration ($p \geq 0.001$). Among the outstanding cultivars in this study were: for TCH V-91-8, V98-120, V99-190, at an average of 73.04; for TAH, V91-8 (9.67), V98-120 (11.21) and V99-190 (9.66), for % POL, CP74-2005, V91-01, C32-368 with average 46.97% for ethanol concentration produced V-91-8, C323-68, CP74-2005 with an average of (46.97%), for (LtEt/hA) the V-98-120,V98-86 and CP74-2005 with an average of 1,717, 29. For efficiency (LtEt/TC) the best cultivars were C32-368 (27.9) V98-6 (55.48) and CP74-2005 (40.84). For production effect in terms of ethanol quality and purity in the municipality of Colón, cultivars V91-8, V99-236 and CP74-2005 should be planted.

Key words: *Saccharum* spp. hybrid; ethanol; variety; sugar fermentation; biofuels.

Introduction

Currently the most important biofuel is ethanol, a 100% renewable product obtained from bioenergy crops and biomass. Fuel ethanol is used to oxygenate gasoline, allowing better oxidation of hydrocarbons and reducing emissions of carbon monoxide, aromatic compounds and volatile organic compounds into the atmosphere. The use of ethyl alcohol as a fuel does not generate a net emission of CO_2 emissions to the environment because the CO_2 produced in the engines during combustion and during the process of obtaining ethanol, is again fixed by the biomass through the process of photosynthesis (Ministry of Agriculture and Breeding (MAC), 1998; Bracho Morán and Labrador Ramírez, 2012).

Among the most widely used bioenergy crops for ethanol production, sugar cane is the most widely used raw material in tropical countries. Sugar cane is a traditional crop in Venezuela; its processing at the level of sugar mills dates back to the 1940s, as well as the beginning of the modernisation and industrialisation of this sector (Aguilar Rivera, 2007; Bracho Morán and Labrador Ramírez, 2012).

The process of obtaining ethanol from sugar cane involves extracting the cane juice (rich in sugars) and conditioning it to make it more assimilable by yeasts during fermentation (Alvarado and El Ayoubi, 2007).

It is a priority to optimise sugar production through producer organisations, specifically for sugar distilleries for ethanol production, and thus promote sustainable economic development with a vision for the expansion and growth of the agricultural and industrial sector in South Lake Maracaibo

(Alvarado and El Ayoubi, 2007; Bracho Morán and Labrador Ramírez, 2012).

All this planning encourages the exploration of new areas with potential for sugar cane cultivation, such as the municipality of Colón in the state of Zulia, introducing a research programme by UNESUR, in agreement with other research support institutions such as INIA (Bracho Morán and Labrador Ramírez, 2012).

In this sense, at the Hacienda La Glorieta experimental field of the UNESUR, Colón municipality, Zulia state, a trial was established with the purpose of evaluating 11 varieties of sugar cane (*Saccharum* spp. hybrid), in the production of ethanol, in a production cycle, which aims to determine the agronomic behaviour and efficiency in the production of sugar and ethanol of the varieties V91-8, V98-86, V91-01, V99-217, C32-368, V98-120, V99-236, B80-408, V00-50, V99-190, CP74-2005, in order to evaluate the best varieties of sugar cane, in the production in tons of sugar cane per hectare (TCH), yield in tonnes of sugar per hectare (TAH), and mainly efficiency in production in litres of ethanol per hectare (LtEt/Ha) and litres of ethanol per tonne of cane (LtEt/TC), the results were analysed with the SPSS statistical package version 2005 for Windows and Tukey's mean comparison test (Bracho Morán and Labrador Ramírez, 2012).

Chapter I. The need for biofuel development in Venezuela: an oil-producing country seeking export diversity

The imminent danger of facing an energy crisis triggered by a sharp increase in international oil prices is currently of great concern and uncertainty due to the disastrous consequences it would generate in several countries that do not have their own natural reserves of fossil fuels (Castro Martínez *et al.*, 2012).

It is not difficult to predict that oil, being a widely used and therefore potentially exhaustible fossil fuel, could significantly diminish its natural reserves in the medium to long term, due to the notable and significant increase in global consumption (Castro Martínez *et al.*, 2012).

Due to its importance, transcendence and topicality, this situation should arouse interest and attention, subjecting the available resources and their energy needs to review and study, seeking to diagnose and mainly evaluate the real feasibility of using alternative renewable energy sources at the national level (Labrador *et al.*, 2008).

Due to this problem, viable alternatives must be sought to minimise the degree of contamination that has been generated by the use of hydrocarbon derivatives. The vegetable agricultural sector represents a parallel alternative for obtaining new sustainable energy sources that could replace traditional sources such as oil and its derivatives in the future (Labrador *et al.*, 2008).

In various countries such as Mexico, Argentina, the United States, Cuba, Colombia and Brazil, the latter being the largest producer and exporter of ethanol in the world, a national plan called E-85 is being carried out, which consists of using fuel for motor vehicles with a composition of 85% ethanol and 15% petrol. In these countries, the transformation of sugar cane juice into ethanol is being evaluated specifically as a complement to be mixed with petrol, without ruling out the possibility that in the future a 100% ethanol-based fuel product could be developed (Amaya *et al.*, 2003).

One of the main benefits of ethanol is the easy transformation of this chemical product based on the derivatives of the sugar cane crop, being one of the crops that covers the largest area of cultivated land at the national level according to Castro Martínez *et al.* (2012), it should only be considered that an abundant amount of biomass is needed to consolidate this transformation process.

It is necessary to establish the cultivation of sugar cane with a proportional increase in sufficient quantity of biomass considering the costs only for the purpose of transformation into ethanol fuel, in order to avoid the deviation of the commercialisation of raw material for sugar to ethanol purposes that could generate a problem among producers due to the fluctuation of prices at the sugar mill or distillation level, thus contributing to the bankruptcy and collapse of the sugar sector (Amaya *et al.*, 2003).

Ethanol production is also seen as an important source of economic income that can be obtained with the technology currently available, and it is also an ecological fuel where the combustion of the product favours the diffusion of few toxic gases into the atmosphere, which would favour environmental impact and less damage to the environment (Bracho Morán and Labrador

Ramírez, 2012).

Research objectives

General objective:

To determine the ethanol yield of 11 varieties of sugar cane (*Saccharum* spp. hybrid) in a production cycle established in the Municipality of Colon, Zulia State (Bracho Morán and Labrador Ramírez, 2012).

Specific objectives:

- Determine the Maturation Curve of the 11 established sugar cane cultivars.
- Calculate the TCH and TAH yields for the 11 established Cane Varieties.
- To evaluate the quality and concentration of ethanol produced per variety.
- Determine the Ethanol Yield per hectare and the efficiency of Lts/Et/Tn of Cane (Bracho Morán and Labrador Ramírez, 2012).

Justification

Humans, like all living beings, depend on the environment for energy. Prior to industrial development, man used animals, plants, wind and water power to obtain the energy necessary for his vital functions, to produce heat, light and transport. Later, man began to use energy sources stored in fossil resources, first coal and later oil and natural gas (Sanhueza, 2009; Bracho Morán and Labrador Ramírez, 2012).

Currently, fossil fuels and nuclear energy provide around 90% of the energy used in the world each year. But fossil fuel reserves are limited and, to a greater or lesser degree, polluting (Bracho Morán and Labrador Ramírez,

2012). Since the mid-20th century, with population growth, the extension of industrial production and the massive use of technologies, concern began to grow about the depletion of oil reserves and environmental deterioration. Since then, the development of alternative energies based on renewable and less polluting natural resources, such as sunlight, tides, water, and bioenergy from biofuels, has been promoted.

In research carried out in some European cities, it has been estimated that 80% of atmospheric pollution is due to the combustion of fossil fuels and that, of this portion, 50% is contributed by transport, with a share of 73.7% of carbon monoxide, 53% of unburned hydrocarbons and 47% of nitrogen monoxide of the total emitted in urban atmospheres. In Central American cities, air pollution is increasing and has already reached dangerous limits for human health and is detrimental to the environment, with motorised vehicles being the main cause of this pollution (Grütter, 1986).

Against this background, ethanol as a biofuel is an alternative renewable energy source whose contribution should not only be measured using energy units, but also considering the powerful multiplier effect generated by the process of transformation and easy development of this energy source from sugar cane cultivation (Poy, 1998; Labrador *et al.*, 2008).

It has been proven that non-polluting plant-based ethanol can be obtained from some agricultural crops that are easy to expand, such as corn, sorghum, grapes, cassava and sugar cane, among others, which can solve the problem of fuel shortages at a given time and halt the deterioration of the environment by using lighter, plant-based fuels (Labrador *et al.*, 2008; Bracho Morán and Labrador Ramírez, 2012).

According to research experiences carried out in countries with experience in the sugar sector such as Brazil, Colombia, Costa Rica and Cuba, it can be mentioned that sugar cane is the main item known in the process of transformation of sugar cane derivatives to produce ethanol. Given that the Southern Lake Zone has the appropriate agro-climatic conditions for the establishment and expansion of sugar cane cultivation, it is feasible to consolidate a research trial to evaluate ethanol yields using various sugar cane cultivars with priority given to mass planting of the crop in order to encourage the potential use of the crop for this purpose (Bracho Morán and Labrador Ramírez, 2012).

By achieving the objectives set out in the project, the expansion of the area cultivated with sugarcane for ethanol can be achieved, technological innovation of the crop will be achieved in the transformation process through the use of simple distillation as a working tool in the final product of ethanol, Given that it will be executed as a field and experimental research at laboratory level, it is necessary to maintain a sequential and meticulous methodology to avoid the minimum possible bias, which allows the collection of sufficient information so that the necessary tools of the analytical statistical programmes of interpretation can be applied to facilitate the result based on the dissemination, dissemination and agricultural extension (Bracho Morán and Labrador Ramírez, 2012).

Depending on the technological advances derived from the process of transforming sugar cane juice into ethanol, it is possible in the area to provide agricultural extension to the producers involved by our illustrious University in the rural area to consolidate the establishment of the planting of this crop for ethanol purposes (Bracho Morán and Labrador Ramírez, 2012).

Feasibility and safety of producing alcohol and other sugar cane by-products in the area south of Lake Maracaibo, as well as self-management of our university in the agrochemical process of producing alcohol as a vegetable fuel (Labrador *et al.*, 2008).

It facilitates the process of technology transfer in research to support other institutions involved in the alcohol issue and will indirectly contribute to the cultural growth and improvement of the quality of life of producers in the area who become involved in this innovative process (Bracho Morán and Labrador Ramírez, 2012).

This technological innovation will improve the infrastructure platform of UNESUR's chemistry laboratories, consolidating the acquisition of the necessary scientific equipment that will contribute to improving the transformation processes in the laboratory area (Bracho Morán and Labrador Ramírez, 2012).

With this research on ethanol production, the first advances are made in terms of publication based on the transformation of sugar cane juice from different varieties of sugar cane sown from certified asexual seed. It will also initiate the process of creating a distillation industry to contribute to the commercialisation of raw materials produced in the area for ethanol production (Bracho Morán and Labrador Ramírez, 2012).

Chapter II. Literature review of the theoretical bases and background to the development of this research.

Labrador Ramírez *et al*, (2020), conducted a trial in the experimental field of UNESUR, Colón municipality, Zulia state, and evaluated eleven sugarcane cultivars (*Saccharum* spp. The criteria evaluated were: production in tonnes of cane per hectare (TCH), tonnes of sugar per hectare (TAH), percentage of Pol (%Pol), litres of ethanol per hectare (LtEt/Ha), product concentration [C], efficiency in litres of ethanol per tonne of cane (LtEt/TC). The results indicated highly significant differences between treatments by variety effects for TCH (Pr >f= 0.0001) TAH (Pr >f= 0.0001) and significant for LtEt/Ha(Pr >f= 0.03) and non-significant differences for % Pol (Pr<f=0.077) ethanol concentration (Pr<f=0.801) and LtEt/TC efficiency (Pr<f=0.621), the best varieties for TCH were V99-236 (177.22), V98-120(181.92), V99-190(189.23), V99-217 (196.04). For TAH, V98-120 (22.7), V99-217(23.72), V99-190 (22.73) and V99-236 (24.1) stood out. For LtEt/Ha the best results were for V99-236 (3,061) and V99-217(2,756.9), for LtEt/TC efficiency the most outstanding cultivars were V99-236 (17.35) and B80-408 (15.28).

Medina (2008), quoted by Bracho Morán and Labrador Ramírez in (2012), carried out a research project at the Universidad del Valle del Momboy, Valera, Trujillo state, Venezuela, with the general objective of studying the technical and economic feasibility of alcohol production in the sugar mills in the state of Trujillo. For this purpose, a projective research was carried out, with a non-experimental field research design, based on the theories of different authors, related to technological innovation and economic technical feasibility of an investment project. The sample consisted of nine (09) workers from three (03) of the best known companies in the panelera industry. A questionnaire containing sixteen (16) questions of the Linker type was used, validated by three (03) experts, with a reliability of Combrach's alpha coefficient of 0.80, processed with descriptive statistics, distribution tables of absolute and relative frequencies. The results indicated few contributions in terms of technological diffusion in products and processes, lack of machinery and technological equipment for these innovations, it was found that the use of technological surveillance and technological foresight is not very widespread in the sugar-panel companies, there are few that maintain continuous development, it was possible to identify the causes that hinder technological innovation in the sugar-panel mills of Trujillo state, the scarce financing from external sources, very high costs, high return periods and very high financial risk for the development of innovation.

Alvarado and El Ayoubi, (2007). They evaluated 13 varieties of sugar cane (*Saccharum* spp. Hybrid) in the experimental field of UNESUR, Zulia state, in Soca II phase for preliminary purposes of ethanol production with the following materials: PR980, PR61-632, V64-10, B67-49, V83-3, V83-8, V83-16, V83-18, V83-21, V83-26, V83-31, CR74-250, RB73-97351. The results indicated non-significant difference in the varieties for TCH (Pr>f=0.1181), highly significant differences for TAH (Pr<f=0.0014), highly significant differences for Lt/Et/ha (Pr<f=0.0001) and for Efic/Lt/Et/Kg Cane, highly significant differences (Pr<f=0.0018); highlighting the best varieties in terms of TCH the V83-3 (134.89), V83-21 (127.19) and V83-18 (124.44) with an average production higher than 124.88 TCH in Soca II phase. For TAH the best yielding varieties were: V83-3 (15.21), V83-

21(11.34), PR61-632 (11.19) with average yield higher than 11 TAH in Soca II phase, with respect to Lt/Et/ha the most outstanding varieties were: CR74-250 (39.767), V83-21 (36.516), PR61-632 (31.432), B67-49 (25.979) and V64-10 (23.919); the efficiency had as the most representative varieties CR7-250 (29.3 Kg/Lt/Et), V83-21 (35 Kg/Lt/Et) and PR61-632 (38.6 kg/Lt/Et).

Díaz *et al.*, (2003), quoted by Bracho Morán and Labrador Ramírez in (2012), carried out a variety trial in the sugar cane zones perimetral to the Turbio river sugar mill, Lara state; where 9 varieties (three controls) of sugar cane were evaluated, in template phase, soca I and soca II, in a randomised block design, 3 replications, 3 rows per plot with 1.5m separation wires. The varieties evaluated were: RB85-5035, RB85-5113, SP70-1284(T), RB85-5546, SP74-2005(T), RB85-5536, C266-70, C137-87 and PR69-2176(T); the variables considered were TCH, TAH and yield. The results of the three cuts indicated that the best varieties in TCH were: RB85-5035 (121), C266-70 (125) and RB85-5113 (114), the lowest values CP74-2005 (92), RB85-5636 (87), PR69-2176 (93). As for the variable TAH, the best results were obtained by: RB85-5113 (13.44), C137-83 (13.42), C266-70 (12.40), being the lowest in TAH: RB85-5536 (9.13) and PR69-2176 (10.04). In terms of yield, the best materials were C137-87 (12.01%), RB85-5113 (11.80%), CP74-2005 (11.24%) and the poorest were C266-70 (9.36%) and RB85-5035 (10.19%). Regarding the three parameters evaluated (TCH, TAH, Yield), the best results were simultaneously reported by the varieties RB85-5113 and C266-70.

Amaya *et al.*, (2003). Researchers of INIA, evaluated in the valleys of Ureña (CAZTA), the group Nº 1 of fundacaña varieties in a randomized block design, three replications, 3 rows per plot, distance between rows 1.5m and 10m linear with 19 varieties in three cycles template, soca I and II, considering the variables ton of cane per hectare (TCH) and tons of sugar per hectare (TAH), the evaluated materials were: B67-49, B74-118, SP70-1284, PR980, PR61-632, SP71-1406, RB85-5546, C85-92, C266-70, C137-81, SP72-4928, RB85-5536, RB85-5035, RB78-4148, CP72-2086, RB85-5113, V71-39, RAGNAR and B81-494. The results during the three cycles indicated that the best varieties in TCH were: B67-49 (140), SP71-1406 (139), SP70-1284 (121), RB78-5148 (132), the lowest results were B81-494 (84), RAGNAR (91) and V71-39 (90). As for the production of tons of sugar per hectare TAH the best materials were: PR980 (12), B74-118 (11.8) RB85-5546 (11.3), while the lowest yielding were: B81-494 (6.8), SP72-4928 (8) and RAGNAR (7.8).

Valecillos (2002). In areas adjacent to the Central Azucarero Venezuela, he carried out a regional trial of 17 varieties of sugar cane in a randomised block design with 3 replications and 3 rows per treatment, in template, soca I and soca II, the varieties sown were: B82-101, V81-1, B82-279, V84-15, V84-8 B82-363, PR61632, V84-9 V8427, V84-2 V78-102, V78-106, V79-101, B82-211, B84-13, V64-10 and B82-12. The parameters evaluated were TCH, tonnes of panela per hectare TPH, yield TCH/TPH, % Pol and purity. The results of the three cycles indicated that the best varieties in terms of TCH were: B82101 (192.32), V84-9 (180.85), V81-1 (174.18) and the most deficient were V79-101 (115.90) V64-10 (112.02) and B82-12 (106.73) in terms of TPH the best varieties were: B82-101 (21.83) V84-15 (21.70) V81-

1 (22.18) B82-12 (13.15) and V84-13 (15.62). In terms of efficiency (TCH/TPH) the best varieties were V79-101 (7.20) V78-102 (7.33) V84-15 (7.75) and PR61-632 (7.76), the poorest were: B82-101 (9.33) V64-10 (9.24) and V84-9 (9.01) for % Pol the best varieties were: V79-101 (13.98%) V78-102 (13.65%) V84-15 (12.90%), the most deficient in Pol being V64-10 (10.88%) B82-101 (10.91%) and V84-9 (10.95%). In terms of purity, the analyses indicated that the best were: PR61-632 (86.92%) V78-102 (86.13%) V79-101 (84.64%) V84-8 (84.97%); the poorest were: B82-101 (77.68%) V84-9 (79.83%) and V64-10 (81.13%).

Theoretical basis

Sugar cane (*Saccharum spp.* hybrid) is a renewable natural resource, because it is a source of sugar, biofuel, fibre, fertiliser and many other products and by-products with ecological sustainability, a cultivar of major importance because 80% of the sugar consumed in the world is produced from it, It is cultivated worldwide and is easy to expand and propagate vegetatively, allows easy agronomic management, botanically sugar cane belongs to the Poaceae family, and there are several species and many cultivars producing sugar, panela and fodder use (Aguilar Rivera, 2007; Sanhueza, 2009; Aguilar Rivera *et al.*, 2012).

The names of the varieties and hybrids consist of a serial number, preceded by the initials of the place of origin. For example: V00-50 means Venezuela, year 2000, lot 50; B80-408 means Barbados, year 1980, lot 408. Each variety has its own characteristics, see table 1 (Bracho Morán and Labrador Ramírez, 2012).

Table 1. Taxonomy of sugar cane (Aguilar Rivera, 2007; Aguilar Rivera *et al.*, 2012).

Kingdom: Plantaae	Division: Magnoliophyta
Class: Liliopsida	Subclass: Commelinidae
Order: Poales	Family: Poaceae
Subfamily: Panicoidea	Tribe: Andropogoneae
Genus: *Saccharum*	Species: hybrid spp.

Morphologically, the sugar cane plant is a crop that adapts very easily to all types of agro-ecological conditions, it is composed of a root system that serves as an anchor for the plant and is the means for the absorption of nutrients and water from the soil, it is made up of two types of roots: roots of the original stake or surface roots, which originate from the root primordium, located in the growth ring of the original piece (stake) that is planted, are thin, highly branched and their life span lasts until the roots appear in the new shoots and occurs between 2 - 3 months of age (permanent roots or support roots). The permanent roots, which sprout from the root growth rings of the new shoots, are numerous, thick and fast-growing, and their proliferation advances with the development of the plant. The quantity, length and age depend on the varieties and environmental factors, such as soil type and humidity, which influence these characteristics (Gómez, 1975; Gravois and Milligan, 1992). In sugar cane, it is difficult to distinguish between superficial roots and support roots, as they accumulate and develop mostly in the first 40 cm of depth (Labrador Ramírez *et al.*, 2020).

The optimum temperature range for sugarcane growth is between 26°C and 30°C, temperatures below 21°C slow down the growth of the stalks and the variation between maximum daytime temperature and minimum nighttime temperature stimulates the concentration of sucrose. The production of biomass in sugar cane is directly related to the solar radiation it intercepts, since the higher the solar radiation, the greater the amount of biomass and the higher the sucrose concentration (Cock *et al.*, 1983; Gravois and Milligan, 1992).

The stalks correspond to the anatomical and structural section of the sugar cane plant, which is of greatest economic value and interest for the manufacture of sugar and the production of alcohol, which is why their chemical composition is of special significance (Cock *et al.*, 1983).

The internode is the portion of the stem, located between two nodes in the apical part of the stem, in the internodes there is the cell division that determines the elongation and the final length. The diameter, colour, shape and length of the internodes depend on the variety, the most common shape of the internodes are cylindrical, barrel-shaped, conoidal, obconoidal and biconcave (Gómez, 1975).

Once the stalk of the stalk is harvested, the roots die, the buds and root primordiums of the vine resprout to give rise to the soca, the number of cuts of the crop (stalk and soca), depends on the variety, the cultural practices and the environmental conditions at the time of harvest; that is to say, there is a tendency to decrease the production as the number of cuts increases (Tecnicaña, 1986).

The leaves of the sugar cane originate at the nodes and are distributed in alternate portions along the stem as it grows, each leaf is composed of leaf lamina, the sheath and the union of these forms the ligule and at its ends are the auricles, which is sometimes pubescent and sometimes glabrous if it lacks hairs. The colour of the ligule and auricle depends on the variety. The leaf lamina is the most important part for the photosynthesis process and its arrangement on the plant depends on the variety, the most common being pendulous or drooping and erect (Gómez, 1975). The arrangement of the leaf lamina on the plant determines yields, and it is possible to find varieties with high or low yields that have different leaf arrangements at any sowing density. The leaf lamina is composed of the midrib, arranged along its length, the secondary ribs parallel to these, the edges with serrated prominences, the number and length of which varies with the variety. The sheath is tubular in shape, wraps around the stem and is wide at the base, it can be glabrous or with stinging hairs in number and length, depending on the variety. The colouring is green when tender, but changes to purplish red when the leaf is fully developed. The intensity of adhesion of the pods to the stem depends on the variety, being preferable the easy detachment, once developed, as it facilitates the burning, the cutting of the plant and reduces the impurities during milling (Tecnicaña, 1995).

The flower in sugar cane has two phases of development, the vegetative phase originated by cell division at the growing points and the reproductive or flowering phase, which is a continuation of the previous one and occurs when the photoperiod, temperature, availability of water and nutrients in the soil are favourable (Labrador Ramírez *et al.*, 2020).

The inflorescence of sugar cane is a silky panicle in the form of a spike. The flower is constituted by a main axis with joints in which the spikelets are inserted, one in front of the other, these contain a hermaphrodite flower with two antennae and an ovary with two stigmas, each flower is surrounded by long pubescences that give it a silky appearance. In each ovary there is an ovule that once fertilised originates the fruit called caryopsis or what is commonly known as the sugar cane seed, which is oval in shape, mm0.5 mm wide and 1.5 mm long (Tecnicaña, 1986).

Chemical composition of sugar cane. In general terms, the chemical composition of sugar cane is the result of the integration and interaction of several factors that intervene directly and indirectly on its contents, varying between lots, localities, regions, climate conditions, varieties, age of the cane, state of maturity of the plantation, degree of stalk tipping, incorporated management, periods of time evaluated, physical-chemical and microbiological characteristics of the soil, degree of humidity (environment and soil), applied fertilisation, among others (Bracho Morán and Labrador Ramírez, 2012).

In global terms, sugarcane consists mainly of juice and fibre, the fibre being the water-insoluble part formed by cellulose, which in turn is composed of simple sugars such as glucose (dextrose). The water-soluble solids expressed as a percentage and represented by sucrose, reducing sugars and other components are commonly referred to as Brix. The ratio between the sucrose content present in the juice and the Brix is called the Juice Purity. The sucrose content, expressed as a % by weight and determined by polarimetry, is known as "Pol" (Labrador Ramírez *et al.*, 2020).

Solar energy is utilised through the plant's photosynthetic (photobiological) conversion mechanism, whereby the plant's own energy from the atmosphere is fixed by the plant into various organic compounds, forming carbohydrates. CO_2 The energy from the atmosphere is fixed by the plant into various compounds of an organic nature, forming carbohydrates. The biomass produced is then transformed into products that have the energetic capacity to substitute petroleum derivatives, such as fuel (anhydrous) alcohol or ethanol (Barreto, 1980; cited by Labrador, 2004).

The plant feedstocks that can potentially be used to produce alcohol are very diverse, although generically they preferentially include those rich in carbohydrates, which can be grouped into two categories from a fermentation point of view:
(a) directly fermentable (glucose, fructose, sucrose)
(b) indirectly fermentable (starch, cellulose)
(Bracho Morán and Labrador Ramírez, 2012; Labrador Ramírez *et al.*, 2020).

According to these categories, the former (directly fermentable) do not require prior transformation into carbohydrates, as is the case with sucrose, glucose and fructose. In the case of indirectly fermentable sources, prior conversion into carbohydrates is necessary, in order to subject them to fermentation so that they can be assimilated by alcoholic yeast, as in the case of starches and cellulose (Labrador *et al.*, 2008).

Although all these carbohydrate sources can be fermented, those with a high concentration of this component in the raw material must be considered initially, which in turn must have high agricultural productivity (t/ha), alcohol yields and profitability (¢/litre). Both starch and cellulose must first be converted (unfolded) into fermentable sugars before being subjected to alcoholic fermentation (Barreto 1980; cited by Labrador, 2004).

Of the cane's composition, 99% corresponds to the elements hydrogen, carbon and oxygen. Its distribution in the stalk is approximately 74.5% water, 25% organic matter and 0.5% minerals (Bracho Morán and Labrador Ramírez, 2012). For many technologists and specialists, sugarcane as a raw material consists mainly of fibre and juice:

CANE = JUICE + FIBRE

CANE = FIBRE + SOLUBLE SOLIDS (BRIX)

Fibre is defined as the fraction of water-insoluble substances that is of interest not only for its quantity but also for its nature, and juice as a dilute, impure sucrose solution. The quality and content of the juice depends to a high degree on the raw material from which it originates. The high % fibre content makes it difficult to extract the juice retained in the cells of the parenchymatous tissue of the stem, which implies and obliges excellent preparation of the raw material for grinding, trying to achieve greater disintegration and rupture of the cells containing the juice (Bennett 1980 cited by Labrador Ramírez et al., 2020).

Soluble solids are represented, as indicated above, by sugars and organic and inorganic non-sugars. Sugars are represented by sucrose, glucose and fructose, the first one having the highest percentage, which can reach values close to 18%. The other sugars in the juice appear in variable proportions, depending on the state of maturity of the raw material. Sucrose is easily hydrolysed in acid solutions according to the following reaction:

$$C_{12}H_{22}O_{11} + H_2O \text{------------------}C_6H_{12}O_6 + C_6H_{12}O_6$$

Sucrose Glucose Fructose

This hydrolytic reaction is generally referred to as inversion, and the monosaccharides, glucose and fructose, produced are called reducing sugars. High levels of these sugars in the stalks indicate a state of immaturity, with the presence of other undesirable substances such as starch. In the case of mature canes, the reducing sugars contribute relatively little to the increased recovery of sugar in the form of crystals (Gravois and Milligan, 1992).

In alcohol production, the use of canes that have not yet reached a satisfactory state of maturity can cause problems, due to the possible presence of undesirable substances for fermentation, since, as indicated, in alcohol production it is the amount of Total Fermentable Sugars (TFA) that is of interest (Alvira *et al.*, 2010).

Indicators of sugar cane production

Maturation index. The time at which the sugar cane crop reaches the point of physiological maturity and is ready for harvesting, determined by the amount of total soluble solids (TSS) present throughout the stalk structure (Alvira *et al.*, 2010).

There are several ways of considering sugar cane maturation, the first is called botanical maturation and is reached when the flower appears, the second is physiological maturation and occurs when the stalk reaches the highest storage of sugar (sucrose) and other solids which can be measured with a refractometer and a third is called economic maturation and corresponds to the moment when the minimum sucrose content is above 13% on a dry weight basis of the sugar cane (Uzcátegui, 1985, cited by Labrador, 2004).

The most common method used is that of physiological maturity, which consists of measuring the amount of sucrose with a hand-held refractometer, the °Brix present in the upper internodes and the °Brix of the lower internodes of the same cane stalk, in order to obtain the ratio between them as an indicator of the degree of maturity (Alvira *et al.*, 2010).

Tons of sugar cane per hectare (TCH) refers to the amount of green cane produced per unit area (ha.) where neither foliage nor root is considered, is a function of yield per variety and represents between 50% and 80% of the total biomass harvested from green cane stalks (Gravois and Milligan, 1992).

Litres of ethanol per hectare (LEH). This refers to the amount of ethanol produced per unit of surface area (ha) and is obtained through the process of extracting sugarcane juice or any other derivative, which is fermented by chemical induction and subjected to a distillation process to separate the alcohol from the water at a temperature of between 75°C and 80°C (Bracho Morán and Labrador Ramírez, 2012).

Alcohol manufacture. Alcohol is made from the fermentation of carbohydrates (sugars or starch), whose raw material will depend on the particular resources and facilities available in each country. Unlike fossil fuels that come from energy stored for long periods in fossil remains, biofuels come from biomass, or organic matter that makes up all living things on the planet. Biomass is a renewable energy source, as its production is much faster than the formation of fossil fuels. Sugar cane is the most suitable source for ethanol production, as the sugars it contains are simple and directly fermentable by yeast (Aro, 2016).

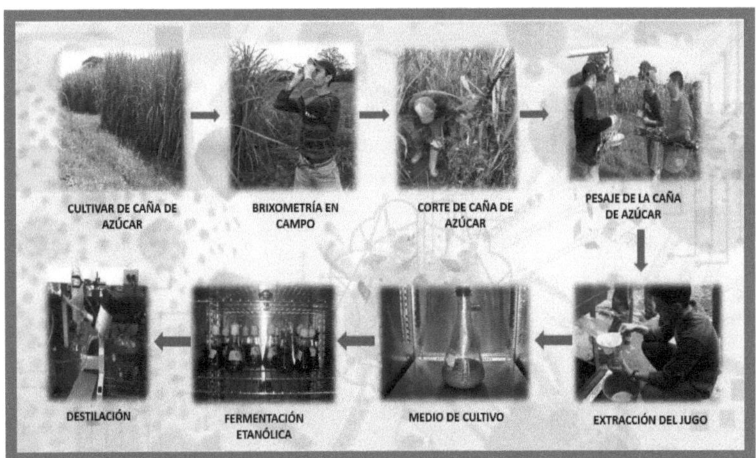

Figure 1. Proposed scheme for obtaining ethanol from sugar cane cultivars at Hacienda La Glorieta, Santa Bárbara de Zulia, Zulia State.

The process of obtaining ethanol from sugar cane involves the extraction of the cane juice (rich in sugars) and its conditioning to make it more assimilable by yeasts during fermentation. From the broth resulting from fermentation, the biomass must be separated to make way for the concentration of ethanol through different unitary operations and its subsequent dehydration, in which it is used as an oxygenating additive, see figure 1 (Aro, 2016).

Hypothesis systems

Alternative hypothesis (Hi). There are differences in ethanol production for the 11 sugar cane varieties evaluated in the UNESUR Experimental Field.

Null hypothesis (H0). There are no differences in ethanol production for the 11 sugar cane varieties evaluated in the UNESUR Experimental Field, some of them showed low ethanol yields.

Variable systems

Table 2. Systems of variables proposed to evaluate the yield for ethanol production in 11 sugar cane cultivars in the southern zone of Lake Maracaibo.

Variables	Description	Units
Dependents	**Ethanol yield**	**LtEt/Ha** **LtEt/TC**
Independent	**The 11 Sugar Cane Cultivars, Sprouting, Ripening, TCH, TAH, Ethanol Concentration**	**Varieties** **%** **°Brix** **Tn**
Interveners	**Soil and climatic conditions**	**mm, °C, %, %, mm, °C**

Chapter III. Methodological framework
Type and design of research. The type of research will be experimental and field research, the quantitative model, sampling of stalks for the field treatments of mature plants to be harvested was carried out (Sabino, 1992; Sabino, 2002). The samples were subjected to a juice extraction process analysed and processed in the laboratory, obtaining ethanol from the fermented sugarcane juice.

The research design was carried out experimentally in completely randomised blocks with three replications and 11 treatments or varieties, the central thread was taken as a sample of 3 kg/repetition each for a total of 99 kilos per treatment/repetition (Bracho Morán and Labrador Ramírez, 2012).

Population and sample. It is represented by plants established in an effective flat area of 3,000 m², with experimental plots of 45 m², divided into three rows of 1.5 metres between rows and 10 metres long, with 6 metres between each block. The overall population comprises 11,880 plants, the eleven varieties evaluated are: V91-8, V98-86, V91-01, V99-217, C323-68, V98-120, V99-236, B80-408, V00-50, V99-190, CP74-2005 (Bracho Morán and Labrador Ramírez, 2012).

The samples considered were 3960 plants from the central wire, taken completely at random from each sugar cane treatment at a rate of three samples/treatment for a total of 99 samples, with a weight of approximately 2-4 kg per sample, to ensure that the results obtained are representative and reliable (Bracho Morán and Labrador Ramírez, 2012).

Research materials and methods. The trial was established at the La Glorieta de la UNESUR farm, Colón municipality, south of Lake Maracaibo, Zulia state, in an agro-ecological zone of tropical rainforest, where the climatic conditions are: average maximum temperature 32.95°C and minimum 22.91°C, average solar radiation 410.24 Mj/ .h, precipitation 1,564.61 mm/year, low cloud cover, maximum relative humidity 92.24%, minimum 53.37%, altitude m4 .s.l. and minimum $53.37\%.m^2$.h, a precipitation of 1.564,61 mm/year, scarce cloudiness, maximum relative humidity 92.24% minimum 53.37%, altitude of m4 .s.n.m and maximum wind speed 8.53 m/sec. Located in the coordinates North Latitude: 08°58'51.9" West Longitude: 71°55'21.9" (Bracho Morán and Labrador Ramírez, 2012).

The research began in June 2008 with the sowing of cuttings of the varieties V91-8, V98-86, V91-01, V99-217, C32-368, V98-120, V99-236, B80-408, V00-50, V99-190, CP74-2005, from the genetic material of the research centre of INIA Táchira, Táchira State. The area of land to be used is 3000 m^2 of flat topography previously prepared with a ploughing pass, two harrow passes and one furrowing pass. m^2 The experimental plot considered will be of 45 m, 1.50 m between furrows and 10 m long, with a separation between the three blocks of 6 m. For the evaluation of the samples, the central furrow of 120 buds per furrow will be considered for a total of 3960 plants (Bracho Morán and Labrador Ramírez, 2012).

The planting of the template (year 2008-2009) was carried out with a density of 12 buds/linear meter of furrow under a completely randomized block experimental design with three replications, 3-row plots and 10 treatments, the trial was conducted by the template cycle; the harvest of the template was carried out after thirteen months (July 2009), according to a schedule of activities. Fertilization in the template phase was carried out according to soil analysis with 280 kg/ha of urea (N H_{24} CO), 200 kg/ha of triple superphosphate and 430 kg of potassium chloride (KCl), applying all the phosphorus 1/3 nitrogen and 1/3 potassium at planting, 1/3 nitrogen and 1/3 potassium at 45 days and the rest of fertilizer at 90 days (Labrador Ramirez *et al.*, 2020).

The soil sample was taken 60 days before sowing and the results revealed (Diaz *et al.*, 2003) 40 cm. deep soil with a clay loam texture, high in phosphorus 46 ppm, high in potassium 212 ppm, medium calcium content 65 ppm, medium organic matter 3.37%. The variables evaluated were the 11 varieties of sugar cane, % sprouting, general appearance of the crop, ripening index and the production variables tons of cane per hectare (TCH), tons of sugar per hectare (TAH), litres of ethanol per hectare (LtEt/Ha), litres of ethanol per ton of cane (LtEt/TC), extraction % yield, % sucrose °Brix. At 45 days after plant establishment, preliminary observations and evaluations of sprouting and general crop appearance of each of the varieties were made. From 10 months (280 days) onwards, the brixometry of the crop was determined every 20 days for the ripening index until 360 days (Bracho Morán and Labrador Ramírez, 2012).

When each of the varieties reached the point of ripening, 10 stalks were sampled from the central furrow, weighed and sent to the laboratory at INIA Yaracuy where they were analysed (%Pol, sucrose, stalk weight, °Brix, among others) (Bracho Morán and Labrador Ramírez, 2012).

Then 3 cane stalks were harvested from the central thread of each treatment and taken to the mill where the juice was extracted by weighing it and measuring its volume, it was passed through a filter to retain coarse foreign particles, then a portion of 1000 mL was taken to make the culture medium in the laboratory, then the materials and the cane juice where the inoculum was formed were sterilised, after which it was left to rest to lower the temperature. The pH was measured and adjusted to 4.5 by adding sulphuric acid ($H_2 SO_4$ 1M) until a pH of 4.5 was obtained.5, then *Saccharomyces cerevisiae* type yeast 10g, 0.65g of calcium sulphate, 0.5g of Urea and small amounts of zinc, cobalt and magnesium sulphate were added, then it was taken to the incubator with a temperature of 32°C and left to rest for 24 hours (see figure 1) (Bracho Morán and Labrador Ramírez, 2012).

Once the culture medium was created, the inoculation was carried out for the 33 samples of 100ml, taking the °Brix of each sugar cane juice, then it was placed in a sterilised 100ml kitasate sealed with a rubber stopper and at the end of the fine mouth a hose was stretched out, resting in a container (Baker) with water to ensure the CO_2 produced by the fermentation process to relieve the pressure of the container, ensuring a completely free environment of O_2 , then the pH was measured in the same way as with the culture medium, adjusting pH 4.5 with sulphuric acid ($H_2 SO_4$ 1M), then inoculated with the culture medium with 3% of the volume of sugar cane juice (3ml), these samples were placed in an incubator at a temperature of 32°C for a period of 5 days (Reyes *et al*, 2022).

Once the fermentation of the samples was reached, the optimum fermentation time was determined (10 to 12 days), then the °Brix of the samples were measured to compare the concentrations of sugars before and after fermentation, then proceeded with the distillation of each fermented juice, placing 100 ml in a distillation balloon (500ml) to install it in the rotary evaporator where the distillation process was carried out, once the thermometer indicates a temperature of 76°C the liquid obtained was discarded, Once the thermometer indicates a temperature of 76°C the liquid obtained is discarded, collecting from 78°C to 92°C where the ethanol plus a small percentage of water will be obtained, then the product obtained from the distillation was measured in a graduated cylinder, then small quantities of ethanol were taken to measure the refractive index with the digital refractometer, After the distillation process was completed, the ethanol yield per Ha (LtEt/Ha) and efficiency (LtEt/TC) were determined for all samples (Bracho Morán and Labrador Ramírez, 2012).

Data collection instrumentation techniques.

- From month 2 after sowing, tests were carried out and samples of % sprouting were taken for each of the varieties sown.

- From the fourth month onwards, the number of millable stems per treatment and general observations on crop adaptation were determined.

- The ripening curve was determined with the refractometer from 280 days after planting, measuring at 20-day intervals, on 5 stems at random from the central furrow of the upper part and then from the lower part (refractometry) of the stem up to 360 days.

- Days before harvesting, samples were taken at random from 10 millable stalks from the central row of each treatment to be analysed in the INIA Yaracuy laboratory to obtain %Pol, bagasse weight and other data.

- Once more than 50% mature, 3 millable stems were harvested per treatment and repetition for the extraction of the juice to be transformed with previous fermentation in ethanol (weighing, fermentation, distillation and the extracted juice). This process was carried out in the Chemistry laboratory of UNESUR, where the ethanol concentration and the LtEt/Ha and LtEt/TC yields were obtained.

- Once all the varieties had matured, harvesting and definitive collection was carried out to measure the tonnes of cane per hectare (TCH).

- The sucrose percentage, purity percentage, sugar content per tonne per hectare (TAH) were determined by laboratory analysis of the samples.

The entire trial evaluation process was recorded in field minutes with tabulated values to be analysed and interpreted (Bracho Morán and Labrador Ramírez, 2012).

Data analysis and processing techniques. The cane was harvested in soca I, determining the weight in tonnes of cane per hectare (TCH) for each of the treatments. The samples were analysed in the chemistry laboratory of UNESUR, by fermenting the cane juice, distilling the fermented juice and measuring the refractive index and volume of the product obtained, in order to evaluate the physical and chemical characteristics of the ethanol. The results for yield, quality and quantity of ethanol produced were then determined. A statistical analysis of variance and Tukey's mean test were used to study the information with the statistical programme SPSS for Windows for the variances TCH, % Pol, TAH, ethanol concentration, LtEt/Ha, LtEt/TC. All the variables analysed will be interpreted by means of histogram figures, interpretation curves, comparisons in the template phase with Soca I and later the results will be interpreted (Bracho Morán and Labrador Ramírez, 2012; Ramírez et al., 2020).

Chapter IV. Results and discussion

The ripening index of the varieties is shown in figure 2, which indicates the harvesting point of the trial that started with the sampling of the° Brix readings. According to these values it can be seen that the varieties V98-120, B80-408, V00-50 and V99-190 behaved as late, since they did not exceed the indicative value of physiological maturity (Pm: 0.95 - 1) even though they were fulfilling the crop cycle and were not suitable for harvesting, since their maturity indicator ranges between 0.7 and 0.9 for the conditions of the municipality of Colón. In order to optimise the yields of the established varieties, harvesting should be carried out at the optimum point of crop ripening, where the high concentration of soluble solids present in the stalk favours yields in sucrose production, contributing to the production of panela.

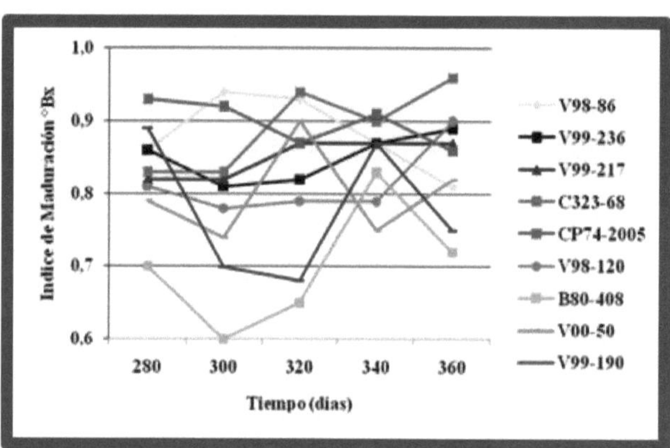

Figure 2. Maturation index of the late varieties under study in eleven sugarcane cultivars (*Saccharum* spp. hybrid) during one production cycle.

Figure 3 shows that the varieties V91-8, V91-01, reached the point of physiological maturation before fulfilling the crop cycle (360 days) behaving as early varieties, because their maturation index exceeds 0.95 value obtained in the brixometry at the moment of the reading in the field, where more than 50% of the materials were at the point of harvesting before fulfilling the crop cycle. In this sense, it can be said that maturation or harvesting point in sugar cane is affected by the genetic characteristics of the variety itself, but restricted by environmental conditions (Valecillos, 2003).

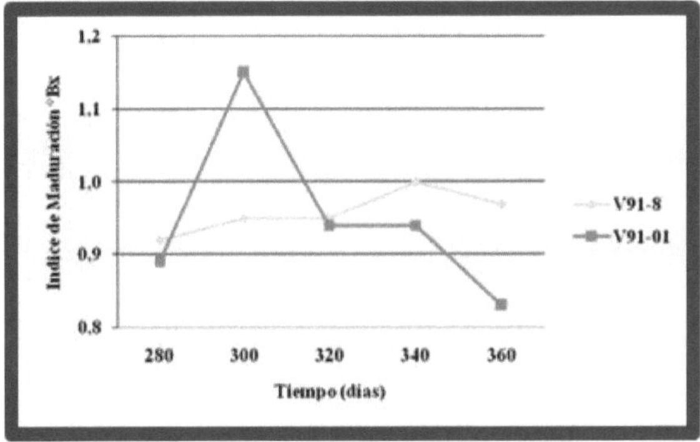

Figure 3. Maturation index of early varieties under study in eleven sugarcane cultivars (*Saccharum* spp. hybrid) during one production cycle.

Figure 4 shows the Tukey test of means for the response to the variable yield in TCH, where the results according to the ANOVA revealed significant differences (Pr>f = 0.0209) for this variable, where three groups of means can be appreciated, the first group conformed by the highest values in TCH for the varieties V91-8, V98-120, V99-190 in an average of 73.04; a second intermediate group conformed by the varieties V91-01, V99-217 and V00-50 with an average value in TCH of 57.17; and a third group with low values in TCH conformed by the materials V98-86, C32-368, V99-236 and CP74-2005 in an average of 35.68.

This response does not exceed the values of the trials conducted by Alvarado and El Ayoubi, in 2007 with the same conditions and with other sugar cultivars, where the average yield was 131.04 for the variable TCH in soca II phase and a non-significant difference for this variable; it should be noted that the yields of this variable in the area were affected by unfavourable environmental conditions for the development of the crop as it presented a period of extreme drought during the critical stage and development of the cultivar (Bracho Morán and Labrador Ramírez, 2012).

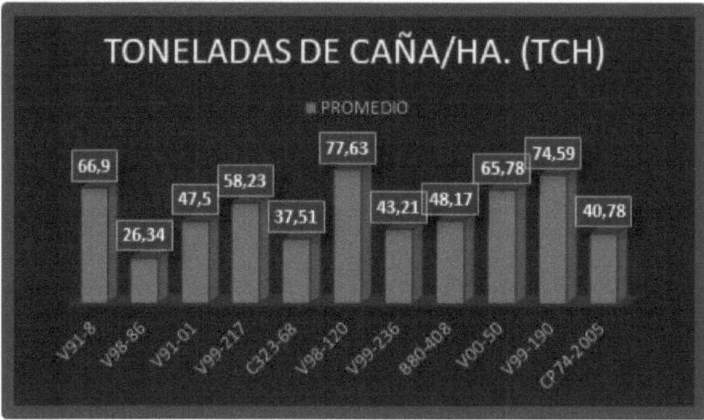

Figure 4. Tons of cane per hectare in eleven sugarcane cultivars (*Saccharum* spp. hybrid) during one production cycle.

Figure 5 shows the results obtained for the Pol percentage variable, where the variety with the best results was: CP74-2005 (54.6%) and the lowest Pol percentage obtained was V99-217 (12.93%). The rest of the varieties studied behaved in a similar way, obtaining values ranging between 20.73 and 42.47%, values considered optimal for an area of high humidity and high rainfall.

Figure 5. Percentage of Pol (sucrose) in sugar cane cultivars (*Saccharum* spp. hybrid) during one production cycle.

Possibly the high % Pol expressed by the varieties was due to good soil fertility and genetic characteristics of the variety, since high % Pol (TSS) is an indicator of high sucrose, sugar and ethanol yields (Gómez, 1975).

Figure 6 shows the averages for the variable TAH where the results indicated two different situations: a first group with sugar production values averaging 10.18 TAH, consisting of V91-8, V98-120, V99-190 and a group with the lowest values averaging 6.11 TAH, consisting of V98-86, C323-68, B80-48.

Tons of sugar per hectare in sugar cane cultivars (*Saccharum* spp. hybrid) during one production cycle at Hacienda La Glorieta, Santa Bárbara de Zulia.

These sugar yields are similar to the trial carried out in optimal zones for sugar cane cultivation in the country, Amaya *et al.*, 2003 in the locality of Ureña, Táchira state, where the best results were for the varieties PR980 (12TAH) and B74-118 (11.8TAH). It is important to highlight that the high sugar yields are due to the high Pol content of the varieties, which can be reflected in the alcohol yields.

Figure 7 shows the response for the variable ethanol concentration where the results revealed a highly significant difference between the materials, differentiating into two groups whose highest values were obtained by the cultivars V91-01, C32-368 and CP74-2005 with an average of 46.97 and a second group with lower values on average of 15.48 in the cultivars V99-217, B80-408 and V00-50.

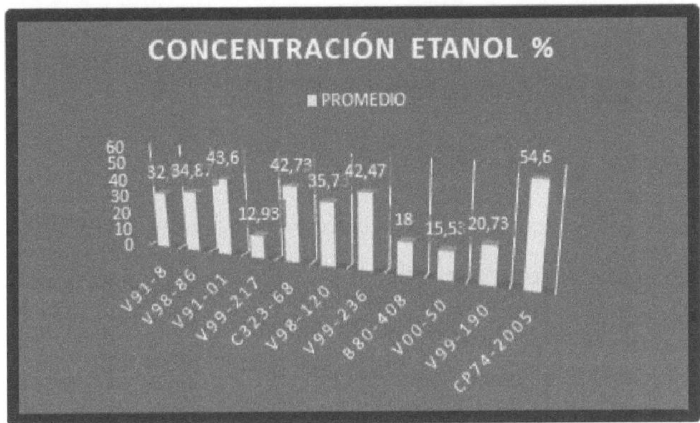

Ethanol concentration (%) in eleven sugarcane cultivars (*Saccharum* spp. hybrid) during one production cycle.

These results exceed the research conducted by Labrador *et al*, 2016 whose averages ranged: V99-236 (31.17%) and B80-408 (24.2%), this is possibly attributed to the refractive indexes obtained for each variety where a low average and in a constant way in their values were obtained, which when interpolated to the SSPS tabulated method of the calibration line (Refractive Index vs % Ethanol) manifest that quality of concentration, i.e. that the ethanol quality efficiency for the 11 varieties are apparently equal and for selection purposes the variety that produced the highest volume of ethanol should be considered.

In the work of Alvarado and Amaya (2021), different techniques were used to make use of lignocellulosic waste such as sugarcane bagasse, through a series of phases that included: pre-treatment of the raw material, enzymatic hydrolysis, fermentation of the sugars by means of yeasts and, finally, the distillation phase to obtain bioethanol. In this study, we worked directly with the raw material and then proceeded to fermentation with *S. cerevisae*, which resulted in an acceptable ethanol production (Figure 1).

Figure 8 shows the response to the efficiency variable (LtEt/TC), for the Tukey mean test, where, according to the analysis of variance, the efficiency was highly significant (Pr>f=0.001). In effect, in the absolute values it can be corroborated that there is a difference, standing out as the best cultivars (V98-86, V99-236 and CP74-2005) that produce more litres of ethanol per tonne of cane.

Efficiency (LtEt/TC) in eleven sugarcane (*Saccharum* spp. hybrid) cultivars during one production cycle.

These results were superior to the research carried out by Labrador *et al.*, 2016, whose response to the efficiency variable (LtEt/TC), for the Tukey mean test indicated non-significant differences; this was due to the similar behaviour obtained by the various cultivars evaluated.

Figure 9 shows the results of the Tukey mean test for the variable LtEt/Ha in which the analysis of variance showed significant differences (Pr>f=0.03) among the varieties that can be attributed to the high contents of % of sucrose, where three groups of means are identified, the most efficient group conformed by the varieties with the highest production in LtEt/Ha: V98-86 V98-120 and CP74-2005 in an average of 1717.29LtEt/Ha, an intermediate group made up of the varieties V91-01, C32-368, V99-236 with an average of 1138.42LtEt/Ha and a third group made up of the most deficient varieties V91-8, B80-408, V00-50 and the V99-190 with an average of 603.82LtEt/Ha. These ethanol volumes obtained are apparently good for the climatic zone of the municipality of Colon due to the high precipitation rates that affect the area.   This response is similar to the results reported by Alvarado and El Ayoubi (2007) for this same variable with different varieties, where they obtained higher ethanol volumes. Similarly, the data obtained were lower compared to the template phase trial conducted by López and Márquez (2009), where the best results obtained for this variable were V99-236 and V99-217 with an average of 2,908.9 LtEt/Ha.

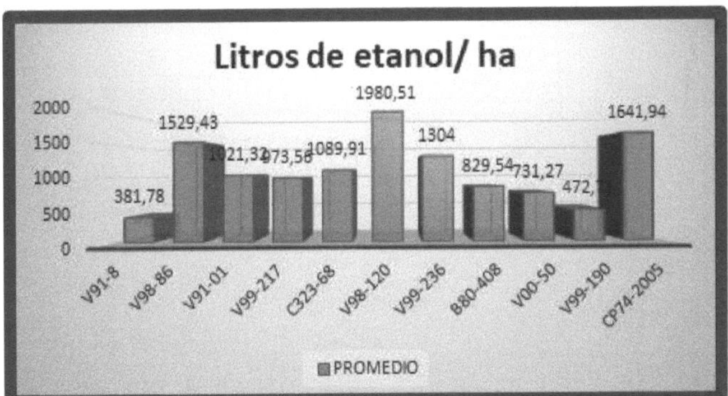

Figure 9. Litres of ethanol per hectare in eleven sugarcane cultivars (*Saccharum* spp. hybrid) during one production cycle at Hacienda La Glorieta, UNESUR, Santa Bárbara de Zulia, Venezuela.

In terms of ethanol production in terms of quality, quantity and purity in the municipality of Colón, the cultivars V91-8, V99-236 and CP74-2005 should be planted, as they offer the best results in this study. On the other hand, if plant production trials are to be established, technological tools must be available that help to reduce the effect of climatic conditions and allow the good development of sugar cane crops. For further studies, it is necessary to increase the number of replicates in order to obtain more precise results regarding the production and quality of ethanol obtained from sugar cane.

Chapter V. Conclusions and recommendations
Conclusions

- The Ripening Index, in most of the varieties was affected by the high rainfall in the area, delaying maturity, however at 360 days half of the trial (2 varieties) showed optimum levels of maturity (V91-8, V91-01).

- The optimum values for tonnes of cane per hectare were expressed by the varieties V91-8, V98-120, V99-190 with an average of 73.04 TCH.

- Excellent values were obtained for yield in % Pol, for an area of high rainfall, where the variety that responded best was the: CP74-2005 with an average of 54.6%.

- The sugar cane varieties that showed the best results were: V91-8, V98-120, V99-190 with an average of 10.18 TAH.

- The results of the production volume in litres of ethanol per hectare the best varieties were V98-86, V98-120, CP74-2005 indicating an acceptable average of 1717.19LtEt/Ha for the area.

- The most efficient varieties in terms of litres of ethanol per tonne of cane are V98-120 (6.43LtEt/TC), B80-408 (6.17LtEt/TC) and CP72-2005 (6.23LtEt/TC).

Recommendations

+ For production effect in terms of ethanol quality and purity in the municipality of Colon, the cultivars V91-8 (15.85%) V99-236 (15.68%) CP74-2005 (54.6%) should be planted as they offer the best concentrations in this trial.

+ In order to establish plant production trials, technological tools must be available that help to reduce the effect of climatic conditions and allow the studies to be carried out to run smoothly.

+ In order to continue the project, it is necessary to increase the number of replicates in order to obtain more accurate results in terms of ethanol production and quality.

References

Aguilar Rivera, N. (2007). Bioethanol from sugar cane. Avances en Investigación Agropecuaria. 11, (3): 25-39. Available at: https://www.redalyc.org/pdf/837/83711303.pdf

Aguilar Rivera N.; Galindo Mendoza, G.; Fortanelli Martínez, J. (2012). Agro-industrial evaluation of sugarcane (*Saccharum officinarum* L.) cultivation using SPOT 5 HRV imagery in Huasteca Mexico. Journal of the Faculty of Agronomy, La Plata. 111 (2): 64-74.

Alvarado Ludeña, G. R.; Amaya Pinos, J. B. (2021). Obtaining bioethanol from sugar cane bagasse by enzymatic hydrolysis. [Degree thesis]. Salesian Polytechnic University. Cuenca, Ecuador. 200 p. Available at: http://dspace.ups.edu.ec/handle/123456789/21229

Alvarado, J.; El Ayoubi, B. (2007). Ethanol: Alternative to obtain fuel from thirteen varieties of sugar cane (*Sacharum* spp. hybrid) in Phase Soca II. [Degree thesis]. Universidad Nacional Experimental Sur del Lago "Jesús María Semprum". Animal Production Engineering. Santa Bárbara de Zulia, Zulia State. 98 p.

Alvira, P., Tomás-Pejó, E., Ballesteros, M., Negro, M. J. (2010). Pretreatment technologies for an efficient bioethanol production process based on enzymatic hydrolysis: A review. Bioresource Technology. 101(13), 4851-4861. https://doi.org/10.1016/j.biortech.2009.11.093

Amaya, F.L.; Hernández, E.Y.; Carrillo, P.; Lindarte, O.; Bonilla, N. (2003). Characterisation of the sugar cane production system in the San Antonio-Ureña Valley, Táchira State, Venezuela. Caña de azúcar. 21 (1): 17-39. Available at: http://sian.inia.gob.ve/canadeazucar/cana2101/arti/amaya_l.htm

Aro, E. M. (2016). From first generation biofuels to advanced solar biofuels. Ambio. 45 (S-1): 24-31. Available at: https://doi.org/10.1007/s13280-015-0730-0

Bracho Morán, N.J.; Labrador Ramírez, J.R. (2012). Evaluation of ethanol yield in 11 sugarcane cultivars (*Saccharum* spp hybrid) in one production cycle. [Special degree work]. National Experimental University of Táchira. Academic Vice-rectorate. Postgraduate Dean's Office. 100 p.

Bull, A. (1969). Efficiency of photosynthesis and respiration in the Calvin cycle and C4 carboxylic acid in plants. Crop. Sci. 9. 726-729.

Castaño, P. and Mejia, G. (2008) Revista de la Facultad de Química Farmacéutica volume 15 number 2, p. 251-258 Universidad de Antioquia Medellín. Colombia.

Castro Martínez, C.; Beltrán Arredondo, L.l.; Ortíz Ojeda, J.C. (2012).

Biodisel or Bioethanol Production A sustainable alternative to the energy crisis? Ra Simhai Journal of Society, Culture and Sustainable Development. 8 (3): 93-100. Available at: chrome-extension://efaidnbmnnnibpcajpcglclefindmkaj/https://uaim.edu.mx/webraximhai/Ej-25barticulosPDF/9%20CASTRO-MARTINEZ.pdf

Cock, J. H.; Luna, C. A. and Palma, A. (1983). Climate and yield in sugar cane. Centro de Investigaciones de la Caña de Azúcar de Colombia (Cenicaña). Technical Series No. 12.

D'Hont A.; Ison, D.; Alix, K. ; Roux, C. ; Glaszmann, J.C. 1998. Determination of basic chromosome numbers in the genus *Saccharum* by physical mapping of ribosomal RNA genes. Genome. 41: 221-225.

Díaz, E; Garrido, A and Anzola, H (2003). National Agricultural Research Institute (INIA). Evaluation of 9 sugar cane varieties. Regional variety trial. Central Azucarero Río Turbio (Lara). X National Meeting of sugar cane varieties. ATAVE, FUNDACAÑA. Guanare, Venezuela. 45-46 pp.

Domínguez, P. S. (1990). Behaviour of the root system of three varieties of sugar cane *Saccharum* spp. in three representative soils of Valle del Cauca. Master's thesis. Faculty of Agricultural Sciences, National University of Colombia. Palmira.

Ferraro, D. (2008). Energy evaluation of ethanol production based on maize grain. Available at: www.scielo.org.ar/scielo.php?pid.

German Agency for Technical Cooperation (UN. ECLAC). (2008). Liquid Biofuels for Transport in Latin America and the Caribbean. ECLAC. 187 p. Available at: https://www.cepal.org/es/publicaciones/3638-biocombustibles-liquidos-transporte-america-latina-caribe

Gómez, F. (1975). Sugar Cane. Edición UPAVE. Centrales Azucareros, C.A. Caracas, Venezuela. 170 p.

Gravois, A. and Milligan, B. (1992). Genetic relationship between fibre and sucrose as yield components. Sci. 32: 62-67.

Grutter, M. 1986. Importance of Atmospheric Pollution. National Autonomous University of Mexico. Faculta de Ciencias. Distrito Federal. Mexico.

Hernández, N. (2007) Presentation to the National Academy of Engineering and Habitat. Accessed 05/08/2009 Available at: www.slideshare.net/energia/etanol.

Hinman, N. (1992). Biomass: An ideal Feedstock for Ethanol Production. Vl 28. California Western Law Review. United State of America.

Humbert, P. (1978). The cultivation of sugar cane. Mexico: Compañía Editorial Continental. 719 pp.

Labrador Ramírez J. R., Razz Garcia, R. C., Bracho Bravo, B. Y., Contreras

Rubio, Q. L. (2020). Ethanol production in 10 sugarcane cultivars *(Saccharum* spp hybrid) in one production cycle. Revista De La Facultad De Agronomía De La Universidad Del Zulia. 36 (2), 111-134. Available at: https://produccioncientificaluz.org/index.php/agronomia/article/view/31198

Labrador, J. (2004). Introduction and evaluation of 13 varieties of sugar cane (Saccharum spp) for sugar cane fodder and sugarcane panel production in the template phase (experimental field of UNESUR), Municipality of Colon, South Lake Maracaibo, Zulia State. Universidad Nacional Experimental Sur del Lago, Academic Direction, Venezuela.

Labrador, J.; Hernández, E.; Amaya F. (2008). Evaluation of 13 varieties of *Saccharum* spp. hybrids for sugar, sugarcane and forage purposes at the seedling stage. Colón Municipality, Zulia State - Venezuela. Agricultural Production. 1 (1): 7 -14.

Labrador, J.; Mora, D.; Alcántara, L.; Paz, F.; Hernández, E.; Contreras, J.; Álvarez, R. (2016). Production of block panela in eleven sugarcane cultivars (*Saccharum* spp. hybrid) in template phase, Colón municipality. Producción Agropecuaria. 5 (1): 3-7.

Laguna Garvett, M. (2011). Agricultural sustainability objectives referring to the trend in ethanol production patterns from corn and sugar cane as feedstock in Venezuela. [Special degree work]. Lisandro Alvarado Central University. Postgraduate in Agrarian Management, Barquisimeto, Venezuela. 246 p. Available at: chrome-extension://efaidnbmnnnibpcajpcglclefindmkaj/http://bibadm.ucla.edu.ve/edocs_baducla/Repositorio/P1221.pdf

Marcano, M.; García, M.; Caraballo, L. (2003). Comparative test of sugar cane varieties in the northeast of Monagas state, Venezuela. Bioagro. 15 (3): 221-225. Retrieved on: 04 July 2024, from http://ve.scielo.org/scielo.php?script=sci_arttext&pid=S1316-33612003000300010&lng=es&tlng=es.

Meléndez, J. R. (2022). Biotechnology and applied management in the production of 1G and 2G bioethanol. Social Science Journal. XXVIII (4): 415-429.

Meléndez, J. R., Mátyás, B., Hena, S., Lowy, D. A., and El Salous, A. (2022). Perspectives in the production of bioethanol: A review of sustainable methods, technologies, and bioprocesses. Renewable and Sustainable Energy Reviews. 160: 112260. https://doi.org/10.1016/j.rser.2022.112260

Meléndez, J. R., Velasquez-Rivera, J., El Salous, A., and Peñalver, A. (2021). Management for the Production of 2G biofuels: Review of the technological and economic scenario. Venezuelan Journal of Management. 26 (93): 78-91. https://doi. org/10.52080/rvg93.07

Ministry of Agriculture and Animal Husbandry (1998). Manual de Cultivos. Caracas, Venezuela.

Monsalve, G. et al. (2006) Revista Dyna November 2006; year 73 Pp. 23 - 27. Accessed 03/07/2009 Available at: http://europa.sim.ucm.es/compludoc/AA?articuloId=592573&donde= castellano&zfr=0

Poy, M. (1998). Does ethanol represent a viable alternative for the sugar cane agroindustry? Rev. Ingenio. Accessed 20/06/2009 Available: http://www.sca.com.co/bajar/Etanol/Mexico/ingenio03.pdf

Reyes Hernández, J.; Torres de los Santos, R.; Hernández Torres, H.; Hernández Robledo E.; Alvarado Ramírez, E.; Joaquín Cancino, S. (2022). Yield and quality of seven sugarcane varieties in El Mante, Tamaulipas. Revista Mexicana De Ciencias Agrícolas. 13 (5): 883-892. Available at: https://doi.org/10.29312/remexca.v13i5.3232.

Sabino, C. (1992). El Proceso de Investigación. Editorial Panamericana, Bogotá, Colombia. 163 p. Available at: chrome-extension://efaidnbmnnnibpcajpcglclefindmkaj/https://www.perio.unl p.edu.ar/tif/wp-content/uploads/2021/04/CarlosSabino-ElProcesoDeInvestigacion_0.pdf

Sabino, C.A. (2002), ¿Cómo hacer una tesis? y elaborar todo tipo de escritos. Editorial Panapo, Caracas, Venezuela. 141 p.

Sanhueza, E. (2009). Agroetanol, an environmentally friendly fuel? Interciencia. 34 (2): 106-112. [cited 2024 Jul 08]. Available from: http://ve.scielo.org/scielo.php?script=sci_arttext&pid=S0378-18442009000200007&lng=es.

Tecnicaña (1986). The cultivation of sugar cane. Editor Carlos Buenaventura. Cali, Colombia: Editorial XYX. 473 pp.

Uzcátegui, C. (1985). Genetic improvement of sugar cane in Venezuela (1962-1982). II selection of introduced varieties. Revista caña de azúcar INIA. Maracay. Maracay. Venezuela.

Valecillos, E. (2002). Evaluation of 10 sugar cane varieties. Second regional variety trial. Central Azucarero Venezuela. X National Meeting of sugar cane varieties. ATAVE, FUNDACAÑA. Guanare, Venezuela. 66-67 p.

Appendices

Figure 10. Distribution of the 11 sugar cane cultivars in the field

Drainage

BLOCK I

T1	T2	T3	T4	T5	T6	T7	T8	T9	T10	T11
V91-8	V98-86	V91-01	V99-217	C323-68	V98-120	V99-236	B80-408	V00-50	V99-190	CP74-2005

BLOCK II

T22	T21	T20	T19	T18	T17	T16	T15	T14	T13	T12
CP74-2005	V98-120	V00-50	V99-190	V99-236	C323-68	B80-408	V91-01	V91-8	V99-217	V98-86

BLOCK III

T23	T24	T25	T26	T27	T28	T29	T30	T31	T32	T33
V99-36	B80-408	V99-17	V98-86	CP74-2005	V91-8	V98-120	V99-190	C323-68	V91-01	V00-50

Figure 11. Average values of^{O} Brix in the field

| NO. | Cultivars | Planting Days | | | | | Average | Crop maturity stage |
| | | 280 | 300 | 320 | 340 | 360 | | |
		I	II	III	IV	V		
1	V91-8	0,9	1,0	1,0	0,9	1,0	0,9	Mature
2	V98-86	0,9	0,9	0,9	0,9	0,8	0,9	Mature
3	V91-01	0,9	0,8	0,8	0,9	0,9	0,9	Mature
4	V99-217	0,9	1,2	0,9	0,9	0,8	1,0	Mature
5	C32-368	0,8	0,8	0,9	0,9	0,9	0,9	Mature
6	V98-120	0,8	0,8	0,9	0,9	1,0	0,9	Mature
7	V99-236	0,8	0,8	0,8	0,8	0,9	0,8	Immature
8	B80-408	0,7	0,6	0,7	0,8	0,7	0,7	Immature
9	V00-50	0,8	0,7	0,9	0,8	0,8	0,8	Immature
10	V99-190	0,9	0,7	0,7	0,9	0,8	0,8	Immature
11	CP74-2005	0,9	0,9	0,9	0,9	0,9	0,9	Mature

INIA - YARACUY
Laboratorio Suelo - Agua - Planta
Resultados de Análisis de Tallos de Caña de Azucar

JM INFORME 24   FECHA:02/07/2010 UBICACION: CAMPO EXPERIMENTAL UNESUR, SANTA BARBARA DEL ZULIA

ISAYO: IREYARUSUE-10-0002 PRODUCCION DE ETANOL A PARTIR DE 11 VARIEDA   RESPONSABLE: JOSE LABRADOR/EDITH

RCELAS 33   GRUPO: CICLO: ESTADO:Zulia   MUNICIPIO JESUS MARIA SA

o rcela	Peso Bagazo	Peso Seco	Peso Tallo (kg)	Brix Corregido (º)	Peso Jugo (g)	Pol (%) Jugo	Fibra Bagazo (%)	Fibra Caña (%)	Pol % Caña	Pureza (%)
1	297	25,6	5,7	17,9	703	14,98	43,80	13,01	12,37	83,69
2	306	25,8	7,8	19,8	694	16,93	43,43	13,29	13,89	85,51
3	347	28,6	4,2	17,5	653	15,01	51,81	17,98	11,41	85,92
4	333	25,2	5,7	13,7	667	10,11	45,18	15,04	8,03	73,96
5	347	27,6	4,5	16,3	653	13,60	49,87	17,30	10,43	83,59
6	379	28,4	6,2	18,0	621	15,45	51,48	19,51	11,33	85,83
7	349	32,8	5,8	17,8	651	15,29	61,94	21,62	11,05	86,04
8	295	27,5	6,2	16,5	705	13,34	49,04	14,47	10,83	81,00
9	334	28,8	5,6	17,0	666	14,34	52,35	17,48	11,03	84,50
10	364	27,0	5,8	15,3	636	12,60	48,99	17,83	9,52	82,51
11	332	25,3	5,5	17,4	668	14,94	43,70	14,51	11,95	86,01
12	329	24,3	6,7	17,9	671	14,93	40,98	13,48	12,11	83,41
13	327	27,2	7,1	15,4	673	12,57	49,09	16,05	9,88	81,78
14	300	25,2	4,4	17,4	700	14,26	43,13	12,94	11,77	82,10
15	369	25,4	4,6	16,7	631	14,20	44,65	16,48	10,89	85,18
16	340	28,9	4,9	16,3	660	13,22	52,90	17,99	10,08	81,25
17	356	27,1	5,6	17,9	644	15,66	48,10	17,12	11,99	87,49
18	315	23,2	4,8	17,3	685	14,87	38,50	12,13	12,33	86,10
19	382	27,0	3,9	16,2	618	13,75	48,81	18,65	10,18	85,03
20	346	29,7	5,0	15,6	654	12,65	55,10	19,06	9,48	81,25
21	395	29,2	5,4	17,7	605	15,23	53,71	21,22	10,81	86,19
22	337	26,0	4,9	19,3	663	17,06	44,55	15,01	13,53	88,39
23	343	24,3	4,9	18,0	657	15,61	41,08	14,09	12,49	86,72
24	300	24,5	5,6	16,7	700	13,48	41,80	12,54	11,18	80,86
25	312	25,5	3,3	15,5	688	12,08	44,89	14,01	9,80	78,09
26	311	24,7	7,6	19,1	689	16,20	41,12	12,79	13,35	84,82
27	352	22,4	4,5	18,1	648	15,97	36,49	12,84	12,93	88,23
28	275	24,5	4,5	16,9	725	13,52	41,44	11,40	11,47	80,14
29	364	23,5	4,2	14,9	636	12,14	40,92	14,89	9,53	81,64
30	385	28,0	3,2	15,0	615	12,43	51,64	19,88	9,04	83,03
31	344	28,4	5,2	18,4	656	15,82	50,94	17,52	12,11	85,98
32	368	28,6	4,1	17,9	632	15,64	51,88	19,09	11,60	87,37

Fecha: 16/03/2010

INIA - YARACUY
Laboratorio Suelo - Agua - Planta
Resultados de Análisis de Tallos de Caña de Azucar

JM INFORME 24   FECHA:02/07/2010 UBICACION: CAMPO EXPERIMENTAL UNESUR, SANTA BARBARA DEL ZULIA

RSAYO: IREYARUSUE-10-0002 PRODUCCION DE ETANOL A PARTIR DE 11 VARIEDA   RESPONSABLE: JOSE LABRADOR/EDITH

RCELAS 33   GRUPO: CICLO: ESTADO:Zulia   MUNICIPIO JESUS MARIA SA

o rcela	Peso Bagazo	Peso Seco	Peso Tallo (kg)	Brix Corregido (º)	Peso Jugo (g)	Pol (%) Jugo	Fibra Bagazo (%)	Fibra Caña (%)	Pol % Caña	Pureza (%)
33	314	28,2	6,2	13,4	686	10,47	52,09	16,36	8,24	78,31

Table 3. Weighing of cane in Kg/plot

No	TREATMENT	REPETITIONS			TOTAL	AVERAGE
		I	II	III		
1	V91-8	424	310	170	904	301,33
2	V98-86	181	60	115	356	118,66
3	V91-01	170	238	234	642	214
4	V99-217	239	298	250	787	262,33
5	C323-68	262	100	145	507	169
6	V98-120	308	451	290	1049	349,66
7	V99-236	149	220	215	584	194,6
8	B80-408	280	225	146	651	217
9	V00-50	385	257	247	889	296,3
10	V99-190	488	340	180	1008	336
11	CP74-2005	248	183	120	551	183,6

Table 4. Tonnes of cane per hectare (TCH)

No	TREATMENT	REPETITIONS			TOTAL	AVERAGE
		I	II	III		
1	V91-8	94,13	68,82	37,74	200,69	66,9
2	V98-86	40,18	13,32	25,53	79,03	26,34
3	V91-01	37,74	52,83	51,94	142,51	47,5
4	V99-217	53,05	66,15	55,5	174,7	58,23
5	C323-68	58,16	22,2	32,19	112,55	37,51
6	V98-120	68,38	100,12	64,38	232,88	77,63
7	V99-236	33,07	48,84	47,73	129,64	43,21
8	B80-408	62,16	49,95	32,41	144,52	48,17
9	V00-50	85,47	57,05	54,83	197,35	65,78
10	V99-190	108,33	75,48	39,96	223,77	74,59
11	CP74-2005	55,06	40,63	26,64	122,33	40,78

Table 5. %Pol (Sucrose)

No	TREATMENT	REPETITIONS			TOTAL	AVERAGE
		I	II	III		
1	V91-8	14,98	14,26	13,52	42,76	14,25
2	V98-86	16,93	14,93	16,2	48,06	16,02
3	V91-01	15,01	14,2	15,64	44,85	14,95
4	V99-217	10,11	12,57	12,08	34,76	11,59
5	C323-68	13,6	15,66	15,82	45,08	15,03
6	V98-120	15,45	15,23	12,14	42,82	14,27
7	V99-236	15,29	14,87	15,61	45,77	15,25
8	B80-408	13,34	13,22	13,48	40,04	13,35
9	V00-50	14,34	12,65	10,47	37,46	12,49
10	V99-190	12,6	13,75	12,43	38,78	12,93
11	CP74-2005	14,94	17,06	15,97	47,97	15,99

Table 6. Tons of sugar per hectare (TAH)

No	TREATMENT	REPETITIONS			TOTAL	AVERAGE
		I	II	III		
1	V91-8	14,1	9,81	5,1	29,01	9,67
2	V98-86	6,8	1,99	4,13	12,92	4,31
3	V91-01	5,66	7,5	8,12	21,28	7,09
4	V99-217	5,36	8,31	6,7	20,37	6,79
5	C323-68	7,9	3,48	5,09	16,47	5,49
6	V98-120	10,56	15,24	7,82	33,62	11,21
7	V99-236	5,05	7,26	7,45	19,76	6,59
8	B80-408	8,29	6,6	4,37	19,26	6,42
9	V00-50	12,26	7,22	5,74	25,22	8,41
10	V99-190	13,65	10,38	4,96	28,99	9,66
11	CP74-2005	8,23	6,93	4,25	19,41	6,47

Table 7. Litres of ethanol per hectare Lt/Et/Ha.

No	TREATMENT	REPETITIONS			TOTAL	AVERAGE
		I	II	III		
1	V91-8	471,11	372	302,22	1145,33	381,78
2	V98-86	2477,66	699,99	1410,65	4588,3	1529,43
3	V91-01	1223,99	1189,99	649,99	3063,97	1021,32
4	V99-217	1104,7	927,1	888,88	2920,68	973,56
5	C323-68	2095,98	799,99	373,77	3269,64	1089,91
6	V98-120	1882,2	2946,5	1063,32	5441,52	1980,51
7	V99-236	1099,27	782,21	2030,54	3912,02	1304
8	B80-408	1088,89	659,99	739,73	2488,61	829,54
9	V00-50	988,16	656,77	548,88	2193,81	731,27
10	V99-190	455,46	498,66	464	1418,12	472,71
11	CP74-2005	1917,85	1951,98	1055,99	4925,82	1641,94

Table 8. Efficiency of litres of ethanol/Tn/Cane

No	TREATMENT	REPETITIONS			TOTAL	AVERAGE
		I	II	III		
1	V91-8	5	5,92	8,01	18,93	6,31
2	V98-86	61,66	52,55	52,25	166,46	55,48
3	V91-01	32,43	22,53	12,51	67,47	22,49
4	V99-217	20,82	14,02	16,02	50,86	16,95
5	C323-68	36,04	36,04	11,61	83,69	27,9
6	V98-120	27,53	29,43	16,52	73,48	24,49
7	V99-236	33,24	16,02	42,54	91,8	30,6
8	B80-408	17,52	13,21	22,82	53,22	17,85
9	V00-50	11,56	11,51	10,01	33,08	11,03
10	V99-190	4,21	6,62	11,62	22,45	7,48
11	CP74-2005	34,83	48,04	39,64	122,51	40,84

Table 9. Concentration %

No	TREATMENT	REPETITIONS			TOTAL	AVERAGE
		I	II	III		
1	V91-8	32,6	31	34,8	98,4	32,8
2	V98-86	34,6	34,2	35,8	104,6	34,87
3	V91-01	50,6	39,8	40,4	130,8	43,6
4	V99-217	14,4	11,8	12,6	38,8	12,93
5	C323-68	50,2	39,2	38,8	128,2	42,73
6	V98-120	32,2	34,4	40,6	107,2	35,73
7	V99-236	52,2	38,6	36,6	127,4	42,47
8	B80-408	14,6	19,2	20,2	54	18
9	V00-50	13	16,6	17	46,6	15,53
10	V99-190	16,6	20,2	25,4	62,2	20,73
11	CP74-2005	54,6	54,4	54,8	163,8	54,6

Brixometry in the field Sugar cane cultivar

Cutting of the Cane   Weighing of the Cane

Juice Extraction Growing Medium

Samples to be Fermented Distillation Process

Results of the statistical analysis Anova Tukey mean tests

⬇ **TCH:**

Obs	trat	bloq	Rend (TCH)
1	1	1	94.13
2	2	1	48.18
3	3	1	37.74
4	4	1	53.05
5	5	1	58.16
6	6	1	68.38
7	7	1	33.07
8	8	1	62.16
9	9	1	85.47
10	10	1	108.33
11	11	1	55.06
12	1	2	68.82
13	2	2	13.32
14	3	2	52.83
15	4	2	66.15
16	5	2	22.20
17	6	2	100.12
18	7	2	48.84
19	8	2	49.95
20	9	2	57.05
21	10	2	75.48
22	11	2	40.63
23	1	3	37.74
24	2	3	25.53
25	3	3	51.94
26	4	3	55.50
27	5	3	32.19
28	6	3	64.38
29	7	3	47.73
30	8	3	32.41
31	9	3	54.83
32	10	3	39.96
33	11	3	26.64

Procedimiento GLM

Información del nivel de clase

Clase	Niveles	Valores
trat	11	1 2 3 4 5 6 7 8 9 10 11
bloq	3	1 2 3

Número de observaciones   33

Litros etanol

Sistema SAS   12:22 Friday, January 22, 2012   1

Obs	trat	bloq	**Etanol**
1	1	1	471.11
2	2	1	2477.66
3	3	1	1223.99
4	4	1	1104.70
5	5	1	2095.98
6	6	1	1882.20
7	7	1	1099.27
8	8	1	1088.89
9	9	1	988.16
10	10	1	455.46
11	11	1	1917.85
12	1	2	372.00
13	2	2	699.99
14	3	2	1189.99
15	4	2	927.10
16	5	2	799.99
17	6	2	2946.50
18	7	2	782.21
19	8	2	659.99
20	9	2	656.77
21	10	2	498.66
22	11	2	1951.98
23	1	3	302.22
24	2	3	1410.65
25	3	3	649.99
26	4	3	888.88

27	5	3	373.77
28	6	3	1063.32
29	7	3	2030.54
30	8	3	739.73
31	9	3	548.88
32	10	3	464.00
33	11	3	1055.9

Sistema SAS   12:22 Friday, January 22, 2012   2

Procedimiento GLM

Información del nivel de clase

Clase	Niveles	Valores
trat	11	1 2 3 4 5 6 7 8 9 10 11
bloq	3	1 2 3

Número de observaciones   33

Sistema SAS   12:22 Friday, January 22, 2012   3

Procedimiento GLM

Variable dependiente: **Etanol**

Fuente	DF	Suma de cuadrados	Cuadrado de la media	F-Valor	Pr > F
Modelo	12	8507932.33	708994.36	2.57	0.0299
Error	20	5515383.14	275769.16		
Total correcto	32	14023315.48			

R-cuadrado	Coef Var	Raiz MSE	etanol Media
0.606699	48.38161	525.1373	1085.407

Fuente	DF	Tipo I SS	Cuadrado de la media	F-Valor	Pr > F
trat	10	7213884.944	721388.494	2.62	**0.0322**

| bloq | 2 | 1294047.390 | 647023.695 | 2.35 | 0.1215 |

Fuente	DF	Tipo III SS	Cuadrado de la media	F-Valor	Pr > F
trat	10	7213884.944	721388.494	2.62	0.0322
bloq	2	1294047.390	647023.695	2.35	0.1215

I want morebooks!

Buy your books fast and straightforward online - at one of world's fastest growing online book stores! Environmentally sound due to Print-on-Demand technologies.

Buy your books online at
www.morebooks.shop

Kaufen Sie Ihre Bücher schnell und unkompliziert online – auf einer der am schnellsten wachsenden Buchhandelsplattformen weltweit! Dank Print-On-Demand umwelt- und ressourcenschonend produzi ert.

Bücher schneller online kaufen
www.morebooks.shop

Printed by Books on Demand GmbH, Norderstedt / Germany